VEGETABLES

Susan Wake

Illustrations by John Yates

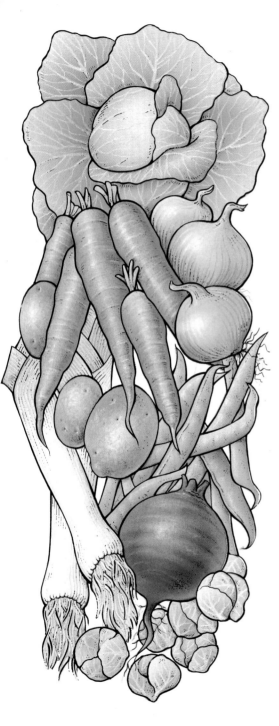

Food

Beans and pulses	**Meat**
Bread	**Milk**
Butter	**Pasta**
Cakes and biscuits	**Potatoes**
Citrus fruit	**Rice**
Eggs	**Tea**
Fish	**Vegetables**

All words that appear in **bold** are explained in the glossary on page 30.

Editor: Fiona Corbridge

First published in 1989 by Wayland (Publishers) Limited
61 Western Road, Hove, East Sussex BN3 1JD, England.

British Library Cataloguing in Publication Data
Wake, Susan
　　Vegetables.
　　1. Vegetables,– For children
　　I. Title II. Series
　　641.3′5

ISBN 1–85210–260–8

Typeset by Kalligraphics Ltd., Horley, Surrey
Printed in Italy by G. Canale C.S.p.A., Turin
Bound by Casterman S.A., Belgium.

Contents

What is a vegetable?

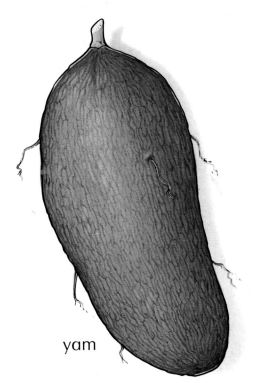

yam

A vegetable is the part of a plant that we eat. It may be the root, stem, leaf, seeds or even the flower and bud. Some of these are eaten raw, and others are cooked.

Some of the vegetables we eat grow wild, but most are specially grown. Vegetables that are easy to grow, such as potatoes, are grown in many countries. Other vegetables, such as

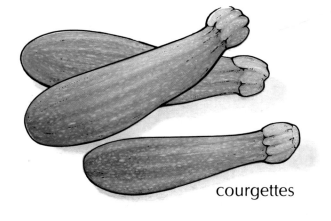

courgettes

pumpkin

4

yams, prefer **tropical climates**. Certain vegetables even grow well in frosty conditions, for example Brussels sprouts.

We now have ways of growing vegetables all over the world, even in countries where they do not grow naturally. Greenhouses allow us to grow plants that only flourish in warm climates. **Irrigation** provides extra water in areas where there is not enough rain. Because of these methods, the vegetables in our shops may come from any part of the world.

sweetcorn

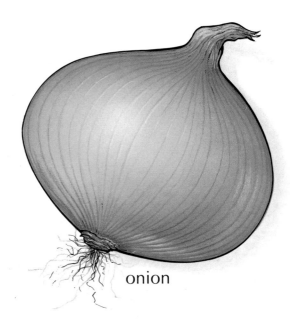

onion

cabbage

5

Different types of vegetables

Root vegetables come from plants that store food in their roots. The roots become fat and swollen, forming the vegetables which we eat. Parsnips and carrots are root vegetables.

Some plants have special underground stems where they store food. These swell to form **tubers**, which we dig up and call vegetables. Potatoes, yams, sweet potatoes and Jerusalem artichokes are all tubers.

We eat the leaves, stems, flowers or buds of some plants. Asparagus and leeks are stems; cabbages and lettuces are leaves. Cauliflowers and broccoli are flower heads, and Brussels sprouts are buds.

Peas and broad beans are seeds. When runner beans and French beans are served, we eat both the seeds and the pods.

There are different varieties of each vegetable.
Next time you go shopping, look at the selection
on sale in your greengrocer's shop.

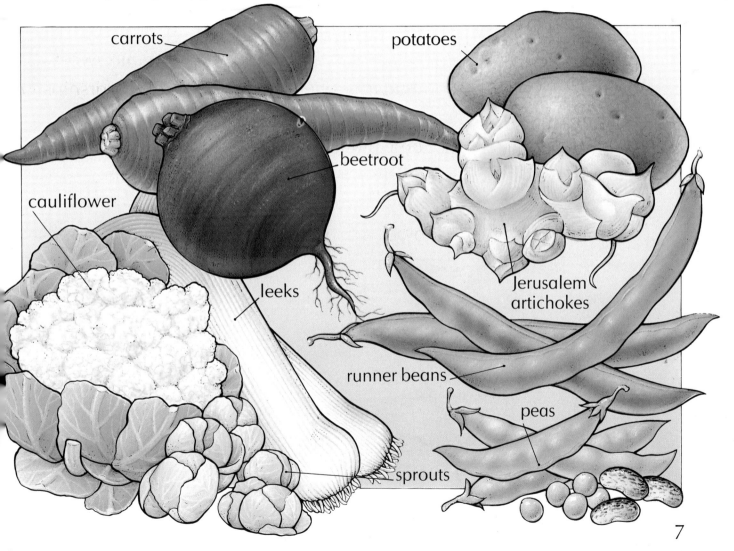

carrots

potatoes

beetroot

cauliflower

Jerusalem
artichokes

leeks

runner beans

peas

sprouts

Fruits or vegetables?

Some of the foods which we call vegetables are really fruits. A fruit is the part of the plant that contains the seeds. If you can see the seeds inside a 'vegetable', it is really a fruit. For example, tomatoes, pumpkins, peppers and

Tomatoes, peppers and cucumbers are all fruits.

Left *Packing tomatoes into boxes on a Moroccan farm, ready to be sent for sale.*

Below *Choosing vegetables from a colourful selection on sale at a market in Bolivia.*

cucumbers are really fruits. We call them vegetables because we tend to think of fruits as sweet and vegetables as savoury. Sweetcorn, too, is not a vegetable: it is a cereal.

Vegetables in the past

Vegetables have been a very important part of our diet for thousands of years. The first humans ate the fruit and roots of wild plants. Later, they collected seeds and planted them to grow food. They made primitive tools to cultivate their plants. Eventually, as people **migrated**, plants were taken to other parts of the world.

Ancient Egyptian tombs have been found to

This illustration shows a medieval farmer ploughing his fields in preparation for planting a vegetable crop.

contain onions and beans, which people believed the dead would need in their next life.

The Romans liked leeks, onions, lettuces, cauliflowers, celery, beans, peas and lupines. Lupines are now used to feed animals. The Romans took their plants to many parts of the world.

In early times, vegetables were not thought to be an important part of a healthy diet, as we know them to be today.

It was not until the beginning of the twentieth century that people realized our bodies need vitamins to keep healthy. Vegetables are a good source of vitamins.

Vegetables for health

It is important to have a balanced diet. We should eat a variety of foods to give us the important **carbohydrates, proteins, fats, minerals** and **vitamins** we need.

Vegetables contain a great deal of water, but are a good source of vitamins and minerals. Our bodies need these in very tiny amounts, but without them we would suffer from diseases.

This diagram shows the approximate amounts of different nutrients contained in vegetables.

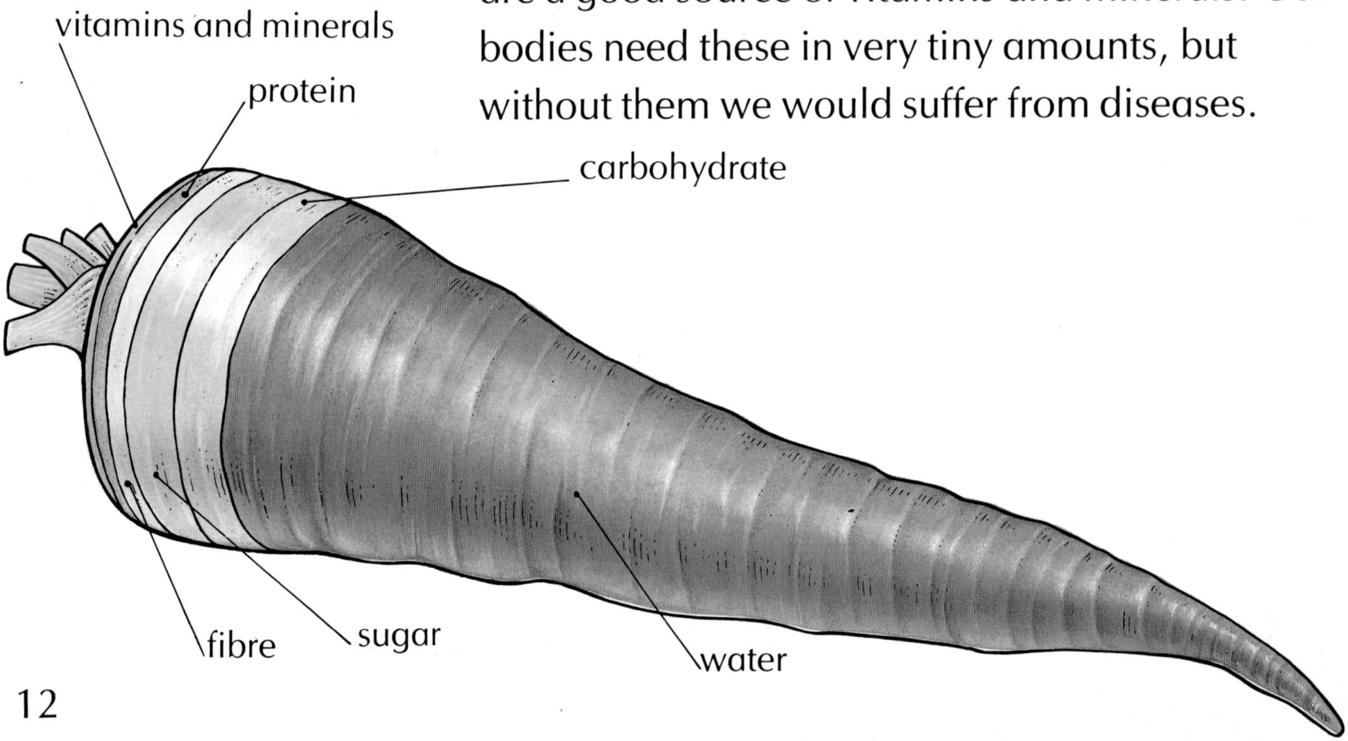

vitamins and minerals

protein

carbohydrate

fibre

sugar

water

Right *Making a salad. Fresh raw vegetables are very nutritious.*

Below *Preparing a meal at a vegetarian restaurant.*

Vegetables also provide **fibre**. This helps our digestive system to work properly.

Vegetables must not be overcooked, or the minerals and vitamins are lost. Raw vegetables, such as we eat in salads, are the most nutritious.

Processes such as freezing, canning and drying can also destroy some of the goodness in vegetables.

Growing vegetables

People grow vegetables in their gardens, on **allotments**, in **market gardens** or on farms.

Plants need **fertile soil**, so vegetable growers add chemical fertilizers or manure to help plants grow well. Growing plants must be protected from harmful pests and diseases. This can be done by spraying with **pesticides** and **fungicides**. Some growers prefer not to use these chemicals.

A farmer working on his vegetable plot in Nanjing, China.

Left *Watering cabbages on a farm in Thailand.*

Below *A farmer sprays his vegetable crop with pesticides. Some farmers prefer not to use these chemicals: their produce is known as 'organic'.*

Greenhouses are widely used in cooler climates to provide warm conditions for seedlings and plants, when the weather is cold. They also create the right temperature for vegetables that normally grow in hot climates.

Farmers sow seeds in batches, which means the vegetables ripen at different times, and there are several harvests. Once harvested, the crops are graded and packed. The produce is transported to the **wholesaler**, who then distributes it to shops, markets and factories.

Vegetable celebrations

Yam festivals are traditional in many parts of Africa and the South Pacific. The best yams are offered to the god of the harvest.

People around the world celebrate the gathering in of the crops they have worked hard to grow, at harvest festivals.

In Europe, corn, vegetables, fruit and sometimes other foods are displayed in churches, and special services are held. The

food is later given to the old and needy.

Some vegetables have become associated with certain traditions. On 31 October in some countries, Hallowe'en celebrations take place. Children often hollow out swedes or pumpkins to make lanterns. In the USA, pumpkin pie has become a traditional part of Thanksgiving celebrations.

Making lanterns out of pumpkins to celebrate Hallowe'en in Ontario, Canada.

Vegetables all year round

When vegetables are harvested, some are sold to be eaten fresh, but often a large part of the crop is kept for eating later in the year. Fresh vegetables will eventually begin to rot, so they must be **preserved**.

Freezing is an important method of preserving vegetables. They must be frozen very quickly.

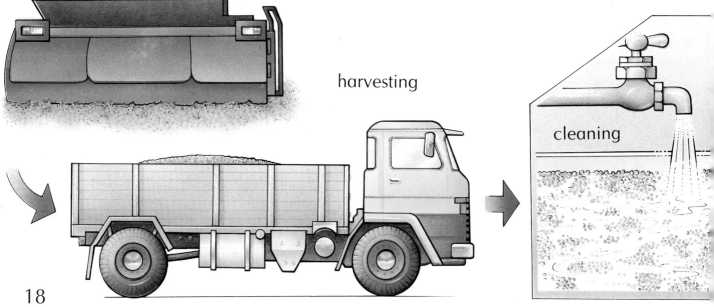

harvesting

cleaning

This is how frozen peas are processed in a factory. The peas are harvested by a machine which shells the peas as they are picked. They are tested for quality, cleaned and frozen. The time from picking to freezing must be no more than 90 minutes!

Drying is an ancient way of preserving food. It is the best method for keeping in the goodness. Vegetables are washed and **blanched** to prevent the loss of colour, texture and nutrients. Hot air is

This diagram shows how frozen peas are processed.

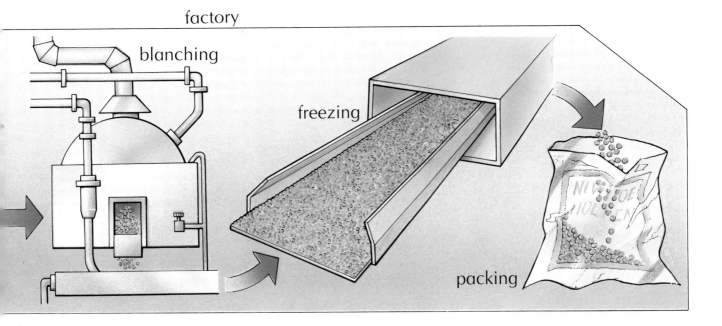

blown over them until they become completely dry. They may go on to become ingredients of dehydrated (dried) soups or meals.

Canning vegetables is another way of preserving them. Some of the goodness is lost, but canned vegetables are a popular and convenient food. The most famous of all canned vegetables is probably baked beans.

In a canning factory, machines fill cans with

Above *Baked beans are one of the most popular canned vegetables.*

Left *Vegetables may be preserved by pickling.*

vegetables. They are made perfectly clean in a **sterilization** tower. The vegetables are then cooked in the cans. Later, the lids are sealed on and the cans are labelled.

Vegetables may also be preserved by being pickled in vinegar. You may have seen this done at home.

These different methods of preserving food mean we are able to enjoy vegetables which may have been harvested many months before.

Modern farming methods, such as this plastic tunnel which provides sheltered conditions, allow vegetables to be grown all year round.

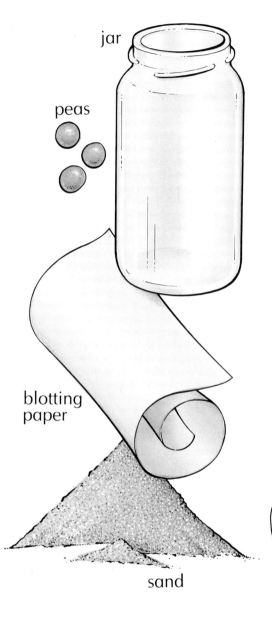

jar

peas

blotting paper

sand

Grow your own peas

You will need: clear jar, blotting paper or kitchen tissue, sand, 2 or 3 pea seeds (which have been soaked for 24 hours), a lolly stick, water.

Put the paper around the inside of the jar. Pack the centre of the jar with sand. Push the soaked seeds down between the sides of the jar and the paper, using a lolly stick. Add water to the sand

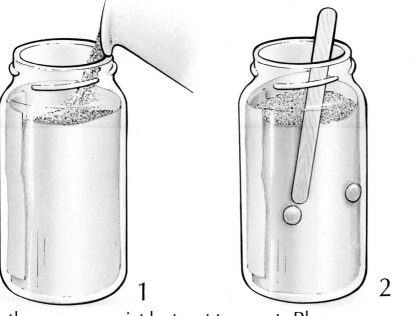

1

2

3

to keep the paper moist but not too wet. Place the jar in a light and warm position.

Observe the progress of the peas every day. Notice if the plumule (young shoot) appears before the radicle (young root). Remember to keep the paper moist with water. Draw your new plant when it has begun to grow.

When the plant looks fairly strong, try planting it into a small pot of soil. Eventually, if it is the right time of year, you can plant it in the garden and produce your very own crop of peas.

23

Vegetables in the kitchen

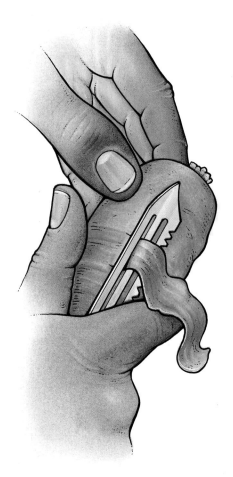

There are many ways of preparing and cooking vegetables. It is important that vegetables are cooked carefully, so that they keep their flavour and nutritional value.

Vegetables should be peeled thinly, as

Above *Peel vegetables thinly.*

Right *A delicious Chinese dish: stir-fried vegetables.*

Left *In India, many people are vegetarian. This man is selling vegetables by the roadside.*

Below *Raw vegetables are the most nutritious. They do not lose any nutrients during cooking.*

nutrients are concentrated just under the skin. You may leave the peel on potatoes and some root vegetables, if you scrub them well. Prepare and cook vegetables just before they are needed, so that they do not lose vitamin C. Cook them with a lid on, until they are only just tender. Overcooking destroys vitamins.

A large part of the world is **vegetarian**. Some people do not eat meat for religious reasons; others because they feel it is wrong to eat animal flesh. Many cannot afford to eat meat.

Russian salad

You will need, for 3–4 people:
1 lettuce heart
225g cooked potatoes
225g cooked carrots
112g cooked peas
112g cooked green beans
mayonnaise
1 large hard-boiled egg
4 gherkins, sliced

1. Wash the lettuce and shake the leaves dry. Arrange them in a salad bowl.

2. Cut the potatoes and carrots into cubes and put them into a large bowl.

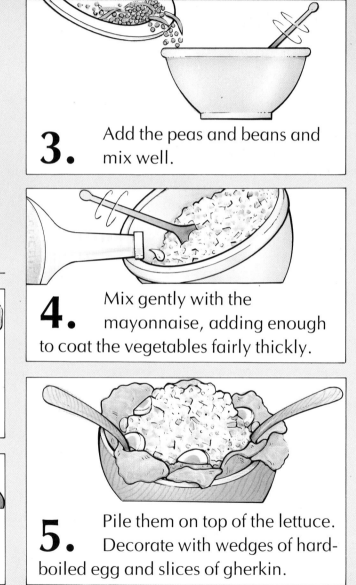

3. Add the peas and beans and mix well.

4. Mix gently with the mayonnaise, adding enough to coat the vegetables fairly thickly.

5. Pile them on top of the lettuce. Decorate with wedges of hard-boiled egg and slices of gherkin.

Sweetcorn fritters

You will need, for 3–4 people:
100g wholemeal flour
1 egg, beaten
150ml milk
200g sweetcorn
oil for frying

1. Put the flour in a bowl. Add the egg and half the milk.

2. Beat until smooth. Gradually add the rest of the milk, beating well, to make a batter.

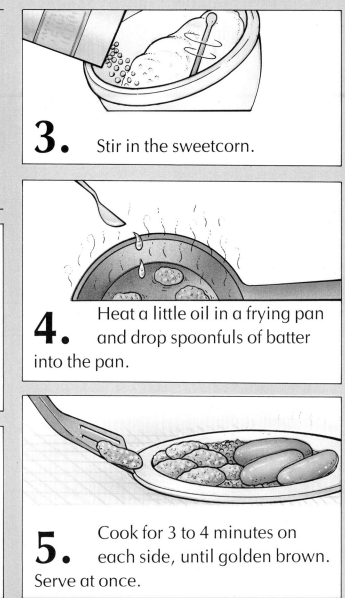

3. Stir in the sweetcorn.

4. Heat a little oil in a frying pan and drop spoonfuls of batter into the pan.

5. Cook for 3 to 4 minutes on each side, until golden brown. Serve at once.

Vegetable broth

You will need:

1 carrot
1 parsnip
half a turnip
1 onion
2 celery stalks
1 leek
1 tablespoon butter
850ml water
1 level tablespoon barley (washed)
1 level teaspoon salt

1. Peel the carrot, parsnip and turnip and cut into small cubes. Wash them well.

2. Chop up the celery and leeks. Wash well. Dice the onion.

3. Melt the butter in a saucepan. Add the vegetables and put the lid on.

4. Fry gently, without browning, for about seven minutes, shaking the pan.

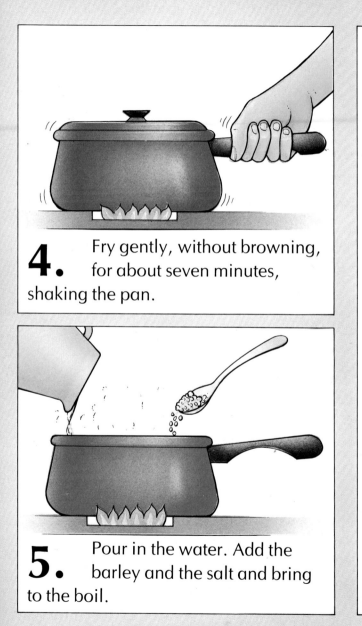

5. Pour in the water. Add the barley and the salt and bring to the boil.

6. Lower the heat and put the lid on. Simmer gently for about one and a half hours or until the barley is soft.

Glossary

Allotment A small amount of land let out for cultivation.

Blanched Put in boiling water for a few minutes.

Carbohydrates Chemical substances found in food, such as starch and sugar which give us energy.

Fats Oily, greasy substances.

Fertile soil Soil which is rich in nutrients to help plants to grow well.

Fibre The coarse part of food that is not digested, but which helps our digestive system to work properly. It is also known as 'roughage'.

Fungicides Substances which destroy fungi.

Irrigation The supply of water by artificial means to allow the growth of food crops.

Market garden A farm specializing in growing many types of vegetables and fruit.

Migrated Moved from one place to another to live.

Minerals Substances, such as iron, which our bodies need in tiny amounts for health and growth.

Nutrients Substances which provide nourishment.

Pesticides Chemicals used to prevent diseases and kill insects and animals which attack plants.

Preserved Kept fresh.

Proteins Any of a large number of food substances essential for healthy growth.

Sterilization The process by which something is made clean and free from bacteria.

Tropical climate A very hot, humid climate found in countries near the equator.

Tubers Swollen underground stems used by certain plants to store food.

Vitamins Any of a number of substances, found in small quantities in food, that are essential for health.

Wholesaler A person who sells things in large quantities to shops or markets.

Books to read

Focus on Vegetables by Andrew Langley (Wayland, 1987)

Food in History by Sheila Robertson (Wayland, 1983)

Growing Food by D.J. Edwards (Rupert Hart-Davis Educational Publications, 1969)

The Food We Eat by Jennifer Cochrane (Macdonald, 1975)

The Vegetable Book by Peter Olney (Puffin, 1979)

Index

Picture acknowledgements

The photographs in this book were provided by: J. Allan Cash 24, 25; H. J. Heinz Company 20 (right); Holt Studios 15 (left); Hutchison Library 9 (right), 14, 16; Christine Osborne 9 (left), 13 (left), 20 (left); Wayland Picture Library 10; Zefa 11, 13 (right), 15 (right), 17, 21.